D1466354

THE FORGOTTEN ARTS

Yesterday's Skills
Adapted to Today's Materials

Book Four

−

Edited by Edie Clark

YANKEE®BOOKS

Camden, Maine

Grateful acknowledgment is made to Luther S. Clark, Jr., for information for the chapter on building bridges.

Illustration Credits:
Chapter One: drawing by Jill Shaffer; photos by Edie Clark. Chapter Three: photo by Debbie Stone. Chapter Four: drawing by Jill Shaffer. Chapter Five: drawings by S. Carl Traegde. Chapter Six: photos by Edie Clark. Chapter Eight: drawings by Jill Shaffer. Chapter Nine: photos by Jean Heavrin. Chapter Ten: drawings by Jill Shaffer.

First Edition 1979

Sixth Printing, 1991

Library of Congress Catalog Card Number 75-10770
ISBN 0-911658-95-5

Table of Contents

The Chimney Connection
Part One

SAFETY IS THE MOST IMPORTANT consideration for the homeowner who is thinking of converting to wood heat. Fire is always a lurking threat — even more so in the country where help is likely to be farther away — if your stove is not properly installed and if your chimney is not in sound condition. The right stove and a solid chimney are the two most important ingredients in safe wood heat. Once you've got these two, then every conceivable precaution must be made to assure the safe installation, operation and maintenance of your wood heat system.

SELECTING A STOVE

Before you select any stove, you should have a clear idea of what the stove's function will be: the amount of space you want to heat, whether or not it will provide all your heat and whether you want it to be decorative, functional or — ideally — a combination of the two. There are more than a thousand models of stoves available at this writing so it is essential to know what you want. Investigate all possible products. Compare construction, costs, efficiency, size of stove in proportion to the living space you need to heat — even looks. Don't hesitate to solicit opinions from other stove owners. Someone who has lived with a particular model of stove usually has a pretty clear idea of the stove's good and bad points.

The abundance of stove models comes from a variety of sources. Established companies have rejuvenated old models, new companies have duplicated once-popular styles or designed new ones, foreign stovemakers, primarily Scandinavian, have been attracted to the booming woodstove market in America and have made substantial contributions to the perfection of the airtight stove, and, of course, older stoves continue to be available.

Be forewarned that although many antique stoves, carefully tended and maintained, have lasted for many years, some of the more recently manufactured stoves probably won't see more than a few seasons of use. They have been hastily designed, poorly constructed and are made of metal that is of questionable quality and thickness. The only real advantage to these stoves is their low price — an immediate savings yet, in comparing prices, you should always be aware that a well-made, cast-iron stove, appropriately cared for, can last you a lifetime.

The older stoves can't compete with the new, airtight stoves in terms of BTUs, but if you have your heart set on an antique stove, you may be lucky enough to find one in mint condition — but be prepared to pay for it. And inspect it very carefully for hairline cracks. You can get a good idea of how "airtight" the old beauty is by dropping a light through the stovepipe opening and darkening the room. Cracks and unnecessary spaces around the door can't hide from the light. Aside from being free from cracks, any stove you select should have all its parts and they should all be in working order. Replacement parts for old stoves are difficult to come by, although sometimes you can find a local welder to restore the stove or manufacture new parts. Another consideration with an old or used stove is the color of the metal: a

stove that has been operated continuously at high temperatures will show a whitish metal discoloration, indicating that improper operation may have weakened the firebox. Although rust will appear on any stove unless it has been carefully tended, this is usually a minor problem and can be eliminated with a wire brush and several coats of stove blacking. Periodic polishing is standard stove maintenance.

If you have an existing flue thimble in your chimney, be sure you know its size before you go stove shopping. Also, be sure you know the measurements from floor to the bottom portion of the flue, an especially important consideration if you're planning to vent the stove into a fireplace, which is lower than many stoves. Some models may have to be eliminated from your list of potentially desirable stoves simply because they won't fit. Your stovepipe must always be either level or running slightly upward, never sloping down into a flue.

Prices for new stoves vary, although all are higher than they were a few years ago. Some people may hesitate to spend as much as the $400 to $800 being asked for some American and many foreign-made stoves. This can be translated into a considerable amount of fuel oil even at today's prices. However, buying a woodstove is not an annual outlay. The cost can be depreciated for as long as the stove is functioning efficiently. Often the more you spend at this initial stage to assure safety, the greater the result in savings — possibly including your house and health. And

7

it's also helpful to keep in mind that, at today's market, stoves do hold their value.

Woodstove stores are mushrooming right along with the manufacturers and a helpful guide to just some of the many new stove models available is Woodstove Directory, 105 West Merrimack Street, P.O. Box 4474, Manchester, NH 03108. This is a 360-page, annually updated directory that costs $3.50. It includes photographs and information such as comments from the manufacturer, stove dimensions, size of firebox, weight, flue size, construction materials, heating capacity and address of manufacturer or distributor.

INSTALLING THE STOVE

Before you buy, you should already have a good idea of where and how your stove will be installed. Until recently many stoves were sold without a word of advice as to how they should be installed and many a homeowner unwittingly took his life in his hands by backing the stove up against a combustible wall, as if it were any other piece of furniture, and plugging the stovepipe into hair-raisingly dangerous chimneys — or worse, sticking the stovepipe out the nearest window. Some early wood heat converts often wonder why they're still alive. Observing the following points will insure a good, safe and efficient installation.

• Stoves are heavy. Before bringing one into the house, check the underpinnings of the floor and add additional support if necessary.

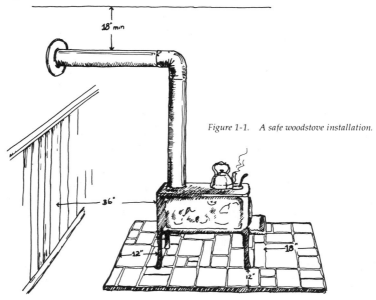

Figure 1-1. A safe woodstove installation.

- Woodstoves are radiant heaters and must be kept at recommended distances from all combustible materials. Wood, wallpaper, many kinds of fabric, most kinds of paint, furniture and even the woodpile itself are all potential fire hazards. A free-standing stove must be 36" from the nearest combustible material in any direction. If gypsum, plaster, asbestos, stone, or brick is used for wall sheathing, most stoves can be brought to within 12" of the wall. If the wall is solid masonry, back the stove right up to it.

- A noncombustible hearth should be provided under the stove and extend 18" out from the firing door, 12" out on either side. This might be a piece of asbestos encased in tin which can be purchased at most hardware stores. Or you can make an attractive hearth out of marble chips, shells, crushed gravel, brick, slate, etc., laid on a metal or asbestos firestop and cut to the proper dimensions.

- Common carbon steel or galvanized stovepipe should be 18" from walls and ceilings. The Shakers often located their woodstoves toward the center of the room and led a well-supported stovepipe at a slight angle under the ceiling to the chimney vent. This method provided added surface for additional heating, but at the same time it increases the hazard of excessive creosote build-up. The longer expanse of pipe permits greater creosote condensation at the cooler end, resulting in greater threat of chimney fires. The shortest length of stovepipe possible is really the wisest installation.

- The firebox should be protected with a bed of fine-grain sand, two inches deep. This layer of sand should always be there to protect the stove bottom from burning out, so be careful not to scoop any of it out when cleaning out ashes or coals.

- Be sure to couple your length of stovepipe so that the fittings project upward to prevent creosote from dripping down the pipe onto your floor. As an additional precaution, you may want to screw each fitting together with self-tapping or pan-head screws. Over a period of time, the subtle vibrations produced by the heat of the stove and movement in the room can loosen these fittings without your noticing it. With the screws in place, this is one less thing for you to worry about. Also, should creosote inside the pipe ignite, it will cause the pipe to vibrate violently, thus uncoupling the stovepipe lengths if they aren't screwed together — and vastly increasing your chances of fully involving the house in a fire, which can remain contained in the chimney.

- Stovepipe should never pass through combustible ceilings or walls unless proper precautions are taken and recommended distances observed. Double- and triple-walled pipe are safe to use and easy to install. Even with a good fire going in the stove, they will only feel slightly warm to the touch. Manufacturers list specifications for clearances. If you don't use a pre-built chimney, cut a hole in the wall or ceiling which provides 18" of clearance around the pipe and fill this hole with noncombustible insulating material or brickwork.

- Try to position the stove as close to the center of your home as possible. Installing a stove at the far end of a wing, for instance, drastically cuts its heating potential. Most new stoves will tell you how many cubic feet you can expect it to heat and you should try to take advantage of as much of that output as you can.

- Do not put the stove near an existing thermostat if your woodstove is to be used for auxiliary heat. This creates an artificial temperature on the thermostat and will hamper the efficiency of the total heat system.

- Be sure to keep open kettles on top of the stove at all times. Wood heat is dry heat and extra moisture is essential.

- To assure good draft, your chimney must rise two feet above the ridge of the roof or two feet higher than any projection within 10 feet of it.

EXCEPTIONAL INSTALLATIONS

Your woodstove should not be vented into the same flue that's being used for either the central heating (oil furnace) vent or a working fireplace. This interrupts good draft and can be hazardous: gases can be drawn downwards and into the house while you are asleep. If you decide to vent the stove through the overmantle of the fireplace, you will be giving up the use of the fireplace. This may be a loss in terms of aesthetics — though true woodstove devotees will argue that point — but it's no loss in terms of heat efficiency. Fireplaces are only about 10 to 20% efficient in their conversion of wood to heat while stoves are from 30 to 80% efficient.

Another possibility to consider, if you are willing to give up the use of an existing fireplace but do not have a flue opening above the mantle, is venting the stovepipe up through the fireplace itself. Forms with a stovepipe collar attached are now commercially available. They will close off the fireplace opening, leaving a thimble for your stovepipe. The form should fit snugly against the damper in your fireplace. If you choose such an arrangement, be sure that the form is attached securely and made airtight by filling all cracks with furnace cement.

If you do not have a chimney at all, or cannot use an existing one safely without extensive renovation, you can beat the problem by installing a new chimney altogether. This can be either masonry (brick, stone, composition blocks) which would require a mason or a metal, pre-built chimney. *Never* try to avoid the expense by venting your stovepipe through a window sash. This merely courts disaster.

Masonry chimneys start from below frost level on a poured concrete footing and are built up; factory-built metal chimneys are supported from the roof and hang down.

Money, time, and labor can be saved by installing a pre-built chimney. These triple-wall (usually stainless steel liner, aluminized steel inner wall and galvanized steel outer wall) chimney pipes are not quite as safe nor as attractive as the masonry chimney but in certain situations, an indispensable innovation. Several nationally known companies manufacture them and provide installation instructions that can be carried out by the average homeowner.

All stovepipes and pre-built chimneys should be inspected period-ically during the heating season. Stovepipe is as essential a link in your wood heat chain as the stove or the chimney. Use 24-gauge stovepipe, if possible. The higher the number, the thinner the metal and some pipe that's available is cheesy and not worth your money. Clean your stovepipe every year, replace it every other year, and keep a special eye on any horizontal sections, where creosote collects. Creosote is highly acidic and corrodes metal. Many stove owners have been dismayed to find the bottom of a horizontal section of pipe corroded through — even while the stove is still in operation.

— *Richard M. Bacon*

Part Two
THE SAFE CHIMNEY

After all the effort of selecting and installing the stove best suited to your needs, be forewarned that your stove is only as good and as safe as the chimney you connect it to. Aside from all the considerations necessary for your own peace of mind, most home insurance policies stipulate that coverage can be suspended if a homeowner knowingly or negligently increases the "hazard of loss" on his property. What this means in the case of a woodstove is that it is your responsibility to observe all local fire ordinances (contact your local fire department) when installing a stove. Should you fail to do so or if the stove is installed in what could be con-

Figure 1-2. The top ten courses of this chimney badly need repointing.

11

sidered a hazardous fashion and a fire occurs that can be attributed to the woodstove, your insurance company will probably not be liable for the damage.

Existing chimneys can often be used to vent a new stove. However, certain checks and precautions must be taken first.

If your chimney is lined and less than fifty years old, your chances are good that a good sweeping or maybe nothing at all is all that's needed. But if the chimney is unlined and old, as was the case with chimneys in many homes built around the turn of the century, you'll probably have some major structural repairs to make. Or the chimney may have to be abandoned or completely rebuilt.

There are several ways you can check the condition of your chimney without the help of an expert. First check the top of the chimney: if the bricks are loose (see figure 1-2), they need to be repointed (this involves removing all the loose bricks down to the level where the

Figure 1-3.   A close-up of a chimney with possibly terminal creosote damage. The acidic creosote has eaten away at the mortar joints, making them porous and a fire hazard.

bricks are solidly in place and rebuilding it from there), a common and relatively inexpensive chimney repair.

Stains on the masonry, resembling long streaks of black ink (see figure 1-3), are telltale signs of mortar that has become too porous, which is dangerous because it allows the flammable creosote to exude from the inside of the chimney out — often to floors or walls which is extremely hazardous in the case of a chimney fire.

Figure 1-4.   Arrows indicate deteriorated mortar that has come loose with the creosote during routine chimney cleaning. This is a sure sign of chimney problems.

Another telltale sign can be found in the clean-out door at the base of the chimney: sift through the soot collected there — if bits of mortar are present (see figure 1-4), this means that the joints are deteriorating and the entire chimney should be inspected.

Another check for deteriorating mortar joints can be done with some helpers, and only if your chimney has some exterior access (what makes a chimney so hard to completely check out is the fact that often interior walls mask part and sometimes nearly all of the chimney, leaving much of its condition to guesswork). Station a helper at each spot where the chimney is exposed — the most common places are the basement and the attic — and seal out as much light as possible. Then shine a bright light up the chimney from the clean-out door at the base. If any of your helpers are able to see even a pinhole of light, this indicates problems.

If all these checks reveal nothing serious, hook up your stove without hesitation. On the other hand, if the chimney appears to have serious structural faults, contact a mason for a repair estimate or else substitute a metal, pre-built chimney. Whatever you do, do it so that no doubt remains in your mind of the chimney's safety.

KEEPING THE CHIMNEY CLEAN

Creosote is one of the less desirable by-products of burning wood and the major cause of chimney fires. Caution should be used in the quality and type of wood burned. Never burn pine or cedar and avoid using very green woods. This is especially important with the new, extremely efficient airtight stoves. These stoves are designed to burn slowly and evenly, reducing stovepipe and chimney temperatures and thus minimizing heat loss. Because of these lower temperatures, the resin in the wood is not completely combusted in the firebox and instead it condenses on the sides of the stovepipe and chimney, forming the flammable soot, creosote. Even

when suitably dry wood is used, there is a steady accumulation of creosote with the airtight stoves and a close watch should always be kept on the condition of the chimney.

A visual inspection of the chimney should be done frequently. Do it during the daylight hours, and a clear day is best. The easiest way is to hold a mirror inside the clean-out door, thus reflecting the interior of the chimney. If you can see a clean square of blue sky reflecting back at you, your chimney is clean — and safe. If, however, the corners of the square are obscured, your chimney needs cleaning. Many variables contribute to how often your chimney will need cleaning — type of stove, frequency of use, condition of chimney, whether or not you keep the dampers closed down or more open, type of wood used and so on — so don't expect a yearly inspection to be adequate.

There are chemicals on the market that can be burned in the firebox to discourage creosote build-up but how effective these are is up for debate. A safer and less expensive way is to clean your chimney yourself when the need arises. There are chimney brushes now on the market that do the very best job of scrubbing the walls of your chimney. These have bristles made of heavy-duty wire or spring steel, are made to fit snugly in standard flue sizes, and act a lot like a bottle brush when used. An alternative is a homemade chimney cleaner that can be rigged fairly simply and, although it won't be as thorough as the commercially available brushes, it will successfully knock down much of the

13

creosote. Make it by wrapping chains or a similar heavy weight in a burlap sack, tie a rope around its neck and raise and lower it against the insides of the chimney walls, being careful not to damage the chimney itself. Be sure you've blocked flue openings and closed all dampers to make sure the soot doesn't permeate the house.

Whether you use the chimney brush or a homemade affair, be forewarned that you're going to get pretty dirty. Recommended attire might be old jeans, a very old sweater or hooded parka, a hat that covers your ears, gloves, dust mask and safety goggles. Those stereotypical depictions of the chimney sweeps of old are no exaggeration. It is grimy.

If you are afraid of heights or prefer not to involve yourself in the truly sooty job of cleaning your chimney, chimney sweeps are coming back in surprising numbers. If a check in your yellow pages does not yield a local sweep, write to The Chimney Sweep Guild, P.O. Box 1078, Merrimack, NH 03054 and chances are they can direct you to a sweep somewhere in your vicinity.

CHIMNEY FIRES

It's important to know something about what it is you're trying to prevent. Chimney fires can be extremely hazardous, especially if they ignite in a chimney that's been neglected, improperly cared for or improperly installed. The dangers of a chimney fire range from burning the entire house down to ruining your chimney to simply getting a free chimney cleaning — but it's not something to be taken lightly or to become overconfident about. Even if your chimney is sound and solid, if it ignites on a windy day, there is still a strong risk that it will ignite your roof.

If there is a build-up of creosote in your chimney, and you are running your stove hot, as is common, for instance, when getting the fire going from morning coals, flames from the firebox reach up the chimney — further than you might expect — and can ignite the flammable creosote. When creosote ignites, its volatile nature, combined with the long, thin passageway of the flue, creates a foreboding roar, something like a jet plane at takeoff, letting you know without doubt that your chimney's on fire. Don't panic. There are reasonable steps that can be taken to bring it under control. But there are a lot of variables, and your particular chimney and stove installation will command individual solutions. You should work out a plan of action in the case of fire ahead of time so you'll know exactly what to do if that unfortunate time occurs. Chimney fires, unlike house fires, are fairly predictable.

Old-timers insist that the last thing you want to do in the case of a chimney fire is to call the fire department. They reason that the firemen will ruin your chimney by hosing it with cold water which will cause the ceramic flue linings to crack. They do have a point in implying that a chimney fire is not quite as catastrophic as some would have you believe but it's also likely that few fire departments still use that method of extinguishing chimney fires. You may want to call your local fire department to ask how they treat chimney fires just for future reference. Whatever the case, if you have a metal pre-built chimney or an unlined masonry chimney, you should call the fire department at once. But, if you're confident of the soundness of your lined, masonry chimney, you don't necessarily have to call for help. As long as it stays contained in the chimney, it's not a threat. Remember that bricks and ceramic flue linings are originally fired at a very high heat and can withstand high temperatures. If the mortar is damaged or if there are cracks in the chimney, flames will seek out the walls adjacent to the chimney and thus involve the house in the fire. And that's the danger: you are not as interested in extinguishing the fire in the flue, as you are trying to prevent it from spreading to the house.

The first thing you want to do is kill the fire in the stove. Do this by throwing sand or coarse salt on the fire or by *gently* spraying a fine mist of water on it — just a *little*. *Never* throw water on a fire in a stove. This is likely to crack or damage the stove and compound your emergency. Next, shut down all the dampers in your stove. You're at an advantage with an airtight stove since dampers control all air intake. There should be no air feeding the flames in the flue. Stoves that are not airtight will still be feeding some air to the fire, through cracks, etc. After you've damped the stove, check the exposed sections of the chimney, usually in the attic. If the bricks are too hot to touch, the chimney is burning dangerously hot and help should be called at once. With an unlined chimney, this might be the case, but with a sound, lined chimney, chances are the bricks will still be amazingly cool in spite of the raging inferno at its core. If you have easy access to your chimney, you may want to block off the top of the burning flue with a solid object such as a steel plate or cement block. But, if your bricks remain cool — someone should stay upstairs throughout, monitoring the heat of the bricks — and if it's not really windy outside, you don't have too much to worry about.

A chimney fire can burn for a while before all the accumulated creosote — in this case, the fuel for the fire — has been exhausted. It burns very hot and very fast, very much like a blowtorch, and the awful roar (and dreadful odor!) can be frightening. But if you keep in mind all the variables and don't let it get blown out of proportion, it can be reasonably controlled and cause no damage whatsoever — and give you a very thorough chimney cleaning at the same time.

If you follow this formula of the right stove, the right wood and a sound chimney, wood heat can be very safe and very satisfying. It requires more input, more time and more energy than the more conventional heating systems most of us grew up with but it's infinitely more rewarding.

— *Edie Clark*

The Community Skating Rink

THERE IS NOT A MORE JOYOUS sight in New England than a small village skating rink aswirl with bright colors and motion on a brilliant winter afternoon. People have been known to move to one of our country towns to be able to drink deeply of the nostalgia stirred by such a scene.

Drifting through memories of the skating days of my childhood — memories of small tots and red tasseled hats, of school-boy hockey and teenagers circling arm in arm, of high snowbanks and frosted breath — I looked out my window one day and realized that an adjoining village green might be a perfect spot to recreate such memories.

Well, as it turned out, perfect was hardly the word. The ensuing four-year saga of battles with the elements, with storms and thaws and dogs and plowmen, with water

commissioners and small children and balky hydrants and road salt and the host of small gremlins that lurk just below the surface of newly created ice, is a tale that needs telling in epic form.

However, this would completely negate my purpose which is to persuade you that, with sufficient courage and tenacity, you, too, can build and manage your own community skating rink. Therefore, we shall proceed in a positive manner with instruction in the seemingly simple art of laying a smooth sheet of ice on the cold winter earth.

You will need:
- A level piece of ground. Do not trust the eye. The simplest sighting device from the hardware store will serve to check the pitch of your proposed site. We built our rink on land that sloped several inches from one edge to

16

the other, and had to fight all winter to maintain a bank of ice at the lower edge which would hold water. Even so, every thaw breached our dike. In the second summer we leveled the rink area with a small bulldozer to leave a broad, faintly dish-shaped field. Seeded over, this is not even noticeable in the summer.

- A source of water. You can pump it from a pond, haul it in with a tank truck, or even use a garden hose on a very small rink, but these are less desirable than a hydrant. If you're near a hydrant, get permission and instruction from your fire department, and persuade them to lend you a 1¼-inch hose, a nozzle, and a combination hydrant wrench and spanner for connecting hose to hydrant and turning on the valve. And if your water is metered, get an okay from the water department to use their water. We pay a token fee each year.

- A willing crew. Eight people divided into four crews has worked well with us. Each crew maintains the rink for one week at a time, usually scraping and watering one night during the week. Our active season (once the rink is built) usually runs from January 1 to early March here in southern New Hampshire. Send the crews a list of their dates, and make a reminder call to each crew at the beginning of its week. Calling an old pal at 10 P.M. on a frigid Friday night to remind him that he and his partner have been remiss all week, and that this is a

great night (it's a nose-grabbing 2° below) to scrape and water for the weekend, turns out to be a shaky way to cement an enduring friendship.

The following are some of the basic construction steps involved in building a sheet of ice.

First, build a base. After pouring thousands of gallons of water into the ground in our first December without a smidgeon of ice to show for it, we discovered that our village was built on a bed of sand. The sand supports our fine old houses admirably, but it soaks up water like a thick diaper under a new baby. We contemplated plastic, but our rink is 65 feet by 85 feet, and plastic is notoriously difficult to handle and prone to leaks. So we came up with an innovation that we think is the *key* to making a skating rink: we *tamp* a snow base.

We tamp every snowfall from early December on. The tamping is done with snowshoes, at least twice over for each snowfall. We have tried rollers and skis and snowmobiles, but for us snowshoes work best.

When we've achieved a packed base of several inches, we water this on cold nights (under 22°) and form a bed of wet slush. The slush freezes solid, though porous at first, and we continue to water lightly on succeeding nights until the frozen slush turns to solid ice. Now, at the next snowstorm, we ask our snowplower to leave walls of packed snow on all sides of the rink, or at least on the low sides, so that we can spray these and build, in effect, a large icy bath-

17

tub that will hold both the water of successive sprayings and the more threatening waters of major thaws. If your site is not close enough to the road to impose upon the plowman, these walls can be shovelled by that willing crew. Be sure to tamp every snowfall once you have started, whether you water or not. One 4-inch fall untamped can harden to the consistency of dry sponge cake and soak up water endlessly without turning to slush.

Starting a rink can be time-consuming enough to warrant a separate crew for this job alone. Once spraying has started it must be done at every available opportunity, even twice a night if need be, to get the most out of a cold spell. A dozen sprayings may be needed to achieve the first good bed of ice. A group of eager and enthusiastic teenagers have proven with us to be a much more ardent, all-weather start-up crew than their shivering elders.

Here are further suggestions:

- Rake the leaves off your rink before snowtime. During our first two seasons we were bedeviled with strange upwellings of soft yellow ice. A laboratory found heavy vegetable matter in the samples we submitted. Now we rake and are plagued no more.

- Water reasonably lightly until you have built a thick layer of ice. Even then, ½ to ¾ of an inch of water per watering is usually enough. A deeper layer carries a lot of heat and can melt a hole in your slab or banking during the night, allowing part of the water

to leak out and leaving your rink a shambles of shell ice.

- Arrange for a dependable plow. One missed snowstorm that turns to rain the next day can leave your rink a lake of virtually unplowable slush.

- Scrape the rink before watering. It's not necessary to sweep. Make a scraper in a form of a T from pine 2x3s, with a handle seven feet long and a "blade" three feet wide. Screw a piece of ¾-inch angle iron to one edge of blade. Brace the handle and sand it smooth.

- Shovel back the edges of the rink to the original boundaries before every watering. Kids knock down the snow banks and the rink rapidly shrinks in size if you don't.

- If you are lower than a nearby street, wall yourself off with a bank of ice or dirt. Salt water and ice are agonizingly incompatible.

- When skaters break through to unaccountably large holes under the ice, pack these with snow, and water with a watering can to make a good patch.

- Hang a light. Kids are home from school at three and it's dark by five.

- Gather your teenagers (and mothers) and decide on separate hours for hockey and free skating. Post these boldly on a large sign. Only if you are a genius at leadership and conciliation will you avoid being eaten alive in battles over this point. Free skating only from 1:00 to 5:00 has worked with us.

And now, if you have not lost your rink two or three times to monumental thaws that were totally unpredicted, if little children have not waded out onto your new partially frozen ice and turned it to lovely shattered glass, if the neighborhood dog pack hasn't repeatedly left its calling cards frozen deeply and darkly into the center of your rink, and if the local garden club isn't ready to sue you for what they are sure you are doing to next year's grass, and even if all these things *have* happened to you, still there was that crisp January morning at 7 A.M. with the dark, frosted trees throwing off the first small splinters of sunlight, when you sailed off all by yourself onto that incredibly smooth gleaming layer of glass, your skate blade crunchily carving the first great circling arc of the day. There was the night the gang came over for a skate in the moonlight and stayed for rum toddies after. And the moment your young daughter completed her first figure eight. And best of all were those Sunday afternoons when you looked out the window to see that colorful whirling mosaic of kids and parents and dogs and visitors, and you said, despite the new gray hairs, it was worth it! Grandma Moses couldn't have painted it prettier.

— *Erik Brown*

19

Drying Fruits and Vegetables

LEANDRE AND GRETCHEN POISSON of Harrisville, New Hampshire, think dehydration is a way of storing food that has been too long neglected. They call it nature's favorite way of preservation, and as far as they're concerned, drying food is more efficient in terms of energy use, nutrition, and aesthetics than canning, freezing, brining, or smoking.

"Dried foods are easy to store as long as they're in a moisture-proof container," Leandre said. "Also, your space requirements are reduced easily by 50 percent, in some cases 90 percent, such as with spinach."

Basically, dehydration is merely eliminating the moisture from the food. The critical factor, however, is the temperature at which the moisture is removed. If the temperature in the drying oven is 140° or 150°, the food may be cooked. Also, if the food is placed in direct sunlight, it may lose its color and vitamins.

"After researching this for two years," he said, "we found that the best drying temperature is between 110 and 120 degrees because that's doing it gently. Temperatures like these don't dry the outside of vegetables and thereby seal in moisture trying to come out. A temperature in this range dehydrates at a rate at which the moisture moves out evenly."

He reminds others that, although dehydration isn't thought of as a familiar preservation method, dried foods are very much part of the general diet and have been for thousands of years. Among other everyday foods, they include grains, spices, herbs, beans, prunes, apricots, raisins, spaghetti, macaroni, even fish and beef.

The next best thing to eating fresh food, Leandre maintains, is eating dehydrated food from a home garden. To this end, he set out to devise a home dehydration method that would dry large quantities of food at a time, such as 20 pounds of carrots, to be competitive with canning and freezing.

He also wanted to have warm air constantly pass over the food, since he found this a better way to dry vegetables and fruits than merely heating them.

What he came up with was a creative way of recycling 50-gallon oil drums: turn them into solar food-driers. His basic design is an oil drum encased in two fiberglass sheets. Aluminum is better than steel for a drier, but better than this is to take an old drum, clean it, cut it, paint it black on the outside, silver inside, and use it instead of letting it rust in a dump.

Applying a nimble imagination, Poisson designed his solar food-drier to take advantage of natural principles. For example, the curved surface of the basic oil-drum design is used as a stationary sunshine collector, whereas a flat collector would have to be moved toward the most efficient angle as the sun travels across the sky. "By positioning this oil-drum design on a north-south axis," Leandre explained, "it sees the same amount of sun all day. You don't have to move it."

In addition, the curved design keeps a constant temperature range inside the drier no matter what the outside temperature is. The system works on the convection principle, that is, when air is moved out of a place, a vacuum is created so that new air moves in.

"The rate of air flow through the drier is determined by the temperature differential," said Poisson. "The bigger the temperature differential, the faster the flow. When the differential is small, the air flow slows down. So what happens in this drier is that if it's a cool day, the air flow adjusts itself. The temperature inside this drier sits between 110 and 120 degrees all the time."

In other words, the drier is designed so that if it's 95° outside, inside it's about 120°. If it's 60° outside, it's 115° inside the drier. Moisture in the air does not affect the operation of the drier itself.

As warm air moves out the exhaust vents, new air comes in the bottom vents of the drier. The air circles around the inside of the solar heat collector and passes now preheated through the food trays. With no moving parts, nothing can wear out.

Leandre and Gretchen harvest their garden and dry their vegetables in the dehydrator. As they suggest to others, they pick the best of the crop at maturity, usually slice the vegetables about one quarter of an inch thick for uniform dehydration, lay the slices on the four screen trays, and set them in the drier. They've dried asparagus, carrots, peas, celery, lettuce, mushrooms, cabbage, squash, onions, peppers, lettuce, basil, and many other vegetables and herbs.

Some foods, such as beans and carrots, are blanched for 30 seconds to fix the color and stop enzyme action. Other vegetables, such as cucumbers and zucchini, don't need blanching. The drying time is usu-

21

ally about one or two days, depending on the moisture content of the individual food.

"You can do things with a dehydrator that you can't do any other way," he said. "For example, zucchini. You can't freeze zucchini unless you like mush. But drying it, zucchini becomes very thin. You can actually use them for chips to dip — they're delicious. When we rehydrate them, they're almost the same as they were originally."

Rehydration is merely adding water back to dried food in much the same way as bringing dried prunes back to full size. Depending on the kind, dried vegetables and fruits are soaked in water from 30 minutes to overnight. Some foods, such as dried spinach, can merely be added by the handful to a soup pot and they quickly spring back to full size and color.

Dried food rehydrates to approximately 90 to 96% of the original size. A bonus of this process is that flavors are concentrated for a richer tasting food because of the 4 to 10% differential in size.

"Herbs you've seen from the store are a brownish color," Leandre said. "That's because they've been dried at too high a temperature, or too fast, or in direct light. All ours remain dark green. People look at them and ask what's wrong with them — they're green."

Storage of dried food must be in moisture-proof jars, which can be any leftover, tightly-sealed peanut butter or mayonnaise jars. One advantage to dehydration is that fewer containers are needed to equal the quantity of food preserved by canning or freezing.

Leandre figures that a single bushel of green peppers takes 16 to 18 quarts to can, 12 quarts to freeze. "That one bushel of peppers, dehydrated, will fit into one quart jar," he said.

Leandre and Gretchen are now at the prototype stage of developing a solar food dehydrator for larger-scale cottage industry. This model, based on the same air-flow concept, is capable of dehydrating 60 pounds of carrots at a time. One model has already been sold to a community canning center in Bath, Maine. The Poissons are recycling old offset printing plates for use as the solar collectors in this design.

One day it dawned on them that Oriental peoples automatically join food preservation with cooking. "The Chinese preserve almost everything by drying," Gretchen said. "They approach their cooking with a little bit of this, a handful of that. The food is reconstituted or stir-fried or put directly into a soup."

The Japanese and Chinese dry their food already cut in the size they're going to be using in their meals. So, using a very sharp Japanese shredder, Gretchen and Leandre began cutting their vegetables to cooking size before drying them.

"In America we're used to lumps or big bowls of vegetables," Gretchen said. "Canning and freezing is the most obvious way of preservation because of the way Americans eat. But because of the way the Japanese and Chinese use food, drying makes a lot of sense."

One fast method of rehydrating food is to put the food in a heavy dish, pour boiling water over it, cover it with a towel, and let the food steep like tea. Dried sliced carrots, for example, can be reconstituted in 10 minutes by this hot-water method.

The Poissons have been very successful in selling the plans for their design, the "Solar Survival Dehydrator," to people in all 50 states, Africa, Europe, South America, Australia, and all the Canadian provinces. (Address: Solar Survival, Box 275, Harrisville, NH 03450.)

"If you're a good scrounge," Leandre said, "it would probably cost about $60 to build the dehydrator yourself. If you had to buy everything and you built it yourself, probably about $100."

The Poissons receive suggestions from people who have built their design. In California someone added two wheels to it and now rolls the food drier around like a wheelbarrow. In Maine blueberry pickers have discovered that if they blanch the berries with wood ash and water, the skin breaks and the berries dry faster.

A year ago they decided to add a backup system for those who wanted or needed a dehydrator in more northern regions where the sun has a shorter day at harvest time. They designed-in three sockets for 75-watt light bulbs in the bottom of the barrel. This way one bulb can be switched on for hazy days. At night all three lighted bulbs can keep the drying process going continuously.

Figure 3-1.   Leandre and Gretchen Poisson harvest a crop of dried vegetables from their solar dehydrator.

The electricity that goes into a simple bulb to create light is relatively inefficient, since much of the energy is wasted as a heat by-product. The Poissons use the heat from the light bulb to boost the efficiency of the food drier.

"The average electric food drier uses about 1200 to 1400 watts of electricity per hour," Leandre noted. "We ran some cost comparisons between electric drying with our three bulbs and with preserving food by canning and freezing. We discovered that using this drier is cheaper than buying just the *lids* to can the food. It's incredible. If you use just the sun, then it's really a big plus. And if you grow food organically, don't use petroleum-based fertilizers, and then dry the food with the sun, you're an absolute plus on the energy scale."

The two of them keep finding new ways to use the drier. Not only can it make yogurt, raise bread dough, dry homemade noodles, make "tempeh," and ripen green tomatoes, but it can also hatch eggs. Gretchen experimented by putting a cookie sheet of water on the bottom level and fertilized eggs on the top. She hatched some chickens this way and figures 300 chickens at one time could be done.

"So now the drier is an incubator," Leandre said. "If it can go 50 miles per hour, we've got it made."

⋘◎⑪◎⋙

Lea's Super Apple Pie

4 cups boiling cider or apple juice
3 cups dried apples
¼ cup maple syrup
¼ teaspoon cinnamon
½ cup seedless Concord grapes or raisins (optional)
1 tablespoon lemon juice
3 tablespoons butter
2 tablespoons flour

Pour boiling cider over apples in a bowl, cover bowl, and let stand 1 hour.

Prepare pastry crust.

Drain apples, reserving ½-cup liquid. Combine drained apples, syrup, cinnamon, flour and grapes or raisins. Place in pastry-lined pie plate. Combine lemon juice and soaking liquid and pour over fruit. Dot with butter. Cover with top crust, cut vents. Bake at 400° for 40 minutes or until apples are tender.

Dried Corn Pudding

1 cup dried corn
2 cups half-and-half, scalded
2 eggs beaten
2 tablespoons dried onions
½ tablespoon dried peppers
1 tablespoon melted butter
Salt and pepper to taste

Place dried sweet corn in half-and-half to soak for 4 hours or overnight in refrigerator.

Put dried peppers and dried onions in a teacup and pour ½ cup boiling water over and let soak 10 minutes. (They will finish reconstituting while they bake.)

Mix the soaked corn in the soaking liquid with the rest of the ingredients and pour into a 1½-quart baking dish.

Bake in a 325° oven 40 minutes or until a knife inserted into the center comes out clean. Makes 6 servings.

— *Steve Sherman*

The Rural Driveway

Today's countryman relies on his driveway as his lifeline to the outside world, especially if his house is set back from a town road — whether paved or graveled. Since by law most New England towns prohibit the use of town equipment on privately owned property, a driveway is the homeowner's responsibility. He must either hire someone with the machinery to build and maintain his drive or do it himself.

Whether using a pick and shovel or engaging a multi-ton bulldozer, he will find the principles of good road construction have not changed appreciably since the days of the Pharaohs. Although scientific knowledge has taught us considerably more about soil structure and its properties since then, the elements of sound construction are the same now as they were thousands of years ago: a firm base and a durable surface. Some variations, improvements, and new materials have been introduced over the years. The Romans, for example, discovered the technique of compacting different layers of construction material to form a firm roadbed. They also realized the importance of good drainage in keeping together a far-flung empire. They scooped up soil from the borders of the proposed bed to create ditches, tamped down the fill on a foundation, and established their throughways above the level of the surrounding ground. Thus they initiated the raised — or high — way, evidence of which can still be found in rural England.

For New Englanders this underlying construction and attention to drainage is especially important because excessive water accumulation coupled with the extremes of seasonal temperatures will tend to

25

play havoc with any road.

Before laying out a good, all-weather driveway or reconditioning an old one, the homesteader should first study the site and try to anticipate some likely problems. Most states have regulations regarding the construction of driveways that give access to state-maintained roads; if you live on such a road check with your state highway department before beginning.

Commonly encountered problems involve having to bridge naturally wet areas, or having to channel surface run-off water to where it can do the least harm to the road, or needing to regulate the pitch of the drive to keep it within acceptable levels. While the logic of "the shortest distance between two points is a straight line" is unarguable, the efficacy of applying this maxim to driveway construction is sometimes questionable. Any of the above problems might be reason enough to reroute a straight-line driveway.

Also, if you live or build on a relatively busy road with a moderate to high speed limit, you may wish to design your drive so that it joins the road with a Y. This will enable a driver to minimize the amount of braking necessary in a fast-moving traffic situation.

In order to plan the most effective drive with the least disturbance to the natural features of the land, it is usually helpful to draw the proposed driveway and its relationship to your house to scale. Thus you can plan a drive wide enough — 9 feet is often cited as the minimum width — and with gradual enough curves to allow easy negotiation. A general

rule of thumb is that a 10-foot-wide driveway can have turns with a radius 35 feet or more, while a 9-foot-wide drive can accommodate turns with a radius of 75 feet or more. Keep in mind that yours will not be the only vehicle to use your driveway. Here in New England you can count on fuel oil trucks — considerably larger than the average family car — making frequent use of your driveway. Take the size of these and other delivery vehicles into consideration when designing turn-around space, and plantings around the perimeter of your driveway. Also, consider snow removal. If possible leave the drive open to any prevailing winds which will blow most of the snow away and make plowing less of a headache and expense. A drive that winds through thick woods has a charm all its own, but be assured that snow will not blow off a woodsy drive when it falls, and will stay later into the spring too.

Take advantage of any natural formations that lie along the route of your drive, perhaps curving the roadway around a large rock outcropping or marshy area, in order to make the drive from the road to house a pleasant transition. If, however, when all factors have been considered a straight drive is the most efficient and easiest to construct, plan a slope that's not too steep if the drive rises or falls from the road to the house. The ideal gradient to aim for is 6%, but as high as 10 to 12% is manageable.

To prevent washouts on steeply-pitched roads — and to avoid having to dig in culverts — many countrymen resort to constructing waterbars or "thank-you-ma'ams" which are low barriers of built-up gravel about one foot wide and extending obliquely across the road from one shoulder to the other. They help to slow down the force of run-off water and divert it from the road surface. Although they are effective both in controlling water and in slowing down traffic, waterbars can be a hazard to the modern driver who is unaware of their existence.

Even if the homesteader is not going to do the actual work himself, he should determine where he wants his driveway, how much he can afford for construction, and be able to communicate these facts to an excavator. Often, if his house is a new one, road work can be done coincidentally with excavating for the cellar.

After marking out the proposed drive, establishing a firm footing is the first step in the actual construction. In New England this excavation must be deeper than it is farther south to combat the freezing and thawing action of the ground in spring. The depth will depend on the terrain and underlying strata but generally should be to about three feet. The subsoil will greatly affect the durability of the finished drive. Coarse-grained, sandy soils provide good drainage whereas fine-grained clay and dense, cohesive soils tend to interfere with proper drainage.

If digging out the roadbed to rid it of unfavorable soil, muck, or organic matter is too much of a problem for the landowner with little equipment, it is best to hire a heavy equipment operator. Although ini-

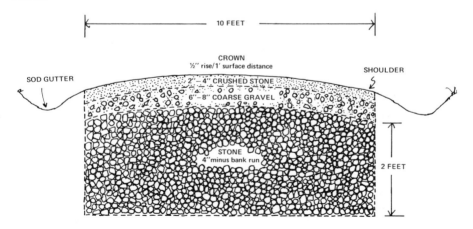

Figure 4-1.   Cross-section of a driveway with good drainage, one that's easily maintained and easily negotiated, even by delivery vehicles.

tially expensive, the work will go quickly and well if the landowner is on hand to emphasize his priorities.

Once excavation has been completed, it is time to plan for future drainage. Sometimes, when the road is to pass through a naturally wet area on somewhat level ground, it will call for installing a culvert. This will lead the water under the roadbed from one side to the other. It can either be the prefabricated, metal type used by highway departments or constructed of rocks, bricks, granite slabs, or reinforced concrete. Whatever the material, the culvert should be set higher at one end than the other — depending on where the water is to go — and braced with large rocks to prevent shifting under the weight of passing cars. It should also be long enough to span the width of the driveway and protrude beyond the shoulders.

Most driveways are crowned unless the rise is gradual over a long distance. This encourages them to shed surface water. The sides of the finished road should rise at the rate of ½ inch from the shoulder to the center line for every foot of surface distance.

To provide for this a foundation layer of coarse stones — called 4″ *minus bank run* by local gravel contractors — can be dumped, spread, and crowned on the roadbed to a depth of about 2 feet. This is most easily done by a bulldozer or flat-bladed grader and followed up with some hand raking. On top of this add a 6- to 8-inch layer of 1½-inch crushed gravel. For best results this should be screened first to eliminate particularly fine particles. Finally,

spread a surface layer of ¾-inch crushed stone to a depth of 2 to 4 inches. Ideally, you should wet this down and roll it to keep it in place.

Sealing the driveway with asphalt or other nonporous material will certainly help to maintain the surface. When applied in semi-liquid form, this seeps through the fine layer of gravel and binds it together. Whether or not to pave your driveway will depend on several factors, and a contractor will have to be retained for this kind of specialized job. One factor is expense; another is anticipating the normal use and weights your road will carry; a third is the surface condition of the road your driveway joins. Many people living in the country choose to construct gravel rather than paved driveways because they believe gravel offers better traction in freezing weather.

Another way to make a durable surface is to pave the driveway with stone blocks or brick. These, however, require hand labor and a tremendous amount of material. To obtain a nearly equally rugged surface, you can spread and roll down a thin layer of breaker dust (a by-product of the gravel process) on top of the crushed stone. Concrete drives leading to a country farmhouse are rare sights.

Once your driveway has been built, further attention should be given to sloping the shoulders. This may require the construction of shallow sod ditches or gutters that parallel the drive and occasionally leave breaks that will cause water to fan out harmlessly in surrounding grassland or woods. Like the principle of the waterbars, these gutters are built to prevent running water — particularly in spring thaws and sudden downpours — from gaining momentum and eroding your drive. These, as well as the culverts, should be inspected periodically to make sure they haven't become clogged with debris (leaves, fallen branches, etc.) that will form natural dams and may feed the water back onto your drive. Sod gutters should be kept shallow enough so as not to encourage too fast a flow.

Even the most carefully con-structed driveway will have to be maintained. Aside from regular inspections, you may have to make sure no waterways form where you don't want them by filling in and raking gullies level. If your drive has been finished with finely crushed gravel, you will want to rake it back from the shoulders occasionally to where it will do more good. And finally, after years of wear, an addi-tional load of gravel should be spread on the surface as part of a regular maintenance program.

— *Richard M. Bacon*

Maple Syrup from Scratch

Y EARS AGO IN THE MID-1930S, when the Depression had its heavy hand on most of us, I worked for a man who harvested maple sap and boiled it down to syrup. The Boss, as he liked to be called, made almost everything he needed for sapping right there on the farm: buckets, spouts, evaporating pans, the fireplace, all handmade from available materials.

Figure 5-2.   A spile, or spout, made from a sumac branch.

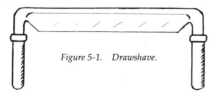

Figure 5-1.   Drawshave.

In the fall the Boss had me cut pieces of ¾-inch red sumac to be made into 9-inch-long spiles, or spouts. We'd peel off the bark while it was still green, then push the pithy center out with a piece of bal-

ing wire, poking away till we had most of it cleared out. The next step was to use a drawshave (see figure 5-1) to whittle off one end. We'd taper down the last three inches on one end of the ½-inch stem to a ⅜-inch diameter at the tip, trying to get an even taper. The spile was finished by whittling off a flat place along one side of the tapered end until an inch or two of the center was showing (see figure 5-2). The Boss preferred his homemade wooden spouts to the metal ones, disliking the way the sap froze on them on cold nights.

Elderberry stems or small birch branches can also be used to make spouts but according to the Boss, elderberry splits too easily when driven into the tap hole. He didn't like birch either because, if driven in too hard, it would split the tap hole in the maple tree.

The Boss also claimed he preferred wooden buckets to galvanized because the wooden ones kept the sap cooler, so it wouldn't be as apt to sour. Birch made the best sap buckets and the Boss made his own, using tools his father and grandfather left him. The wood came from birch logs (white, yellow, red, or black) cut from his woodlot and taken to a small sawmill nearby. The Boss let these 1x4 boards season under cover for a year or more. After they'd dried, he'd cut some into 13-inch lengths. Using an old sap bucket stave as a template, he'd mark the outline for a new stave. The marked board was 3¾ inches at the top end, tapering on both sides to 3½ inches at the bottom end. This would be the outside of the new stave, so the Boss always made sure any flaws or blemishes would be on this side of the new board.

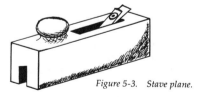

Figure 5-3. Stave plane.

Next, this marked board would be put into a vise, lengthwise, pencil-marked side toward him, and he'd straddle the board with his "stave" plane (see figure 5-3), keeping the

high side of the bevel closest to him. Thus he cut a beveled edge (108°) to the pencil line, then turned up the other edge and planed the other side the same way.

Next, laying the board flat with the inner face up, the Boss would mark a line ¾ of an inch up from the bottom end (the smaller end), then put another line ½ an inch up from the first line. Using a wooden mallet and wood chisel, he'd cut a V notch across the bottom of the stave between the two lines. He notched into the center slightly deeper than at the edges, cutting the notch about halfway into the stave. Ten of these finished staves (see figure 5-4) were enough to make one sap bucket.

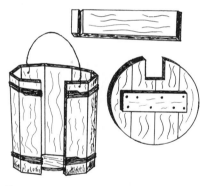

Figure 5-4. Sap bucket, with one stave removed, and a cover.

Three more things were needed to make a sap bucket: the bottom, hoops, and bail, or handle. The bottom was made by laying four, 12-inch 1x4 boards edge to edge. A 10-inch 1x4 cleat was used to fasten these four boards together, centering the cleat both ways of the

31

boards, and using 1½-inch screws in holes that had been pre-drilled to prevent the cleat from splitting. The Boss drove a small finish nail partway into the center of the cleated boards, then with a pencil and a 6-inch piece of string, he drew a 12-inch circle. He shortened the string to 5½ inches and drew an 11-inch circle inside the 12-inch one. Another 11-inch circle was made on the inside bottom of the bucket before he turned the bottom cleat-side up again, and cut along the 12-inch circle line with a compass saw. (I bet the Boss would have liked to have had a sabre saw for times like this.) He finished the bottom by tapering both sides from the 11-inch pencil marks to the center of the 12-inch edge, using a wood rasp. There was more rasping till 10 staves and the bottom fit snugly.

Next were the hoops. The Boss put 10 staves around the bottom piece and held them there temporarily by winding baling wire around the middle of the bucket. He made me fetch a pail from near the small pot-bellied stove that heated the workshop. Coiled around in the pail were 7-foot-long pieces of ½-inch-thick elm branches. He'd had them soaking in warm water for several hours. He would lay the wired bucket on its side and start the first hoop around, about an inch above where the bottom fitted into the staves. The flat side of the elm was laid onto the stave and a couple of small copper nails were put in. The next stave in the bucket was turned up and the hoop was nailed to it. The Boss repeated this until he had a little more than two full turns of the

elm nailed to the bucket. He put another hoop around the bucket, about 2 inches down from the top, the same way. He removed the temporary baling wire when the elm hoops had dried. Then a doubled piece of baling wire, long enough to make a handle, was fastened into small holes made at the top of the bucket. A sketch of the completed bucket is shown in figure 5-4.

The Boss made the buckets watertight by filling them with water and letting them soak for a few days, but I've heard since that soaking wooden sap buckets in hot glycerine (200° F.) for several hours, then storing them away for a summer will also make them watertight. It might be a good idea to rinse them with water before using them, to keep from getting too much glycerine in the sap. Painting the outside of the sap buckets white might be good to help preserve the wood and keep the buckets from being warmed too much by the sun.

We used covers on our sap buckets to keep out the snow, rain, and dirt. Four 1x4s, each 15 inches long, were cleated together with a 14-inch 1x4, then the Boss cut this into a 15-inch circle. A cut in the edge, 4 inches in and 4 inches wide, formed the opening where the sap dripped into the bucket. A completed cover is shown in the sketch in figure 5-4.

I never saw the Boss make an evaporating pan. The two he had were made and used quite a few years before I went to work on his place. But one day when the sap wasn't running very well, he told me how he made the pans. Using a sheet of copper, 5 foot square and ⅛-

inch thick, he marked a line 6 inches in on all four sides of the sheet. A 6-inch cut was made on each of the four corners, as shown in figure 5-5. This strip was bent up 90° along each side, to form the four sides of what was now a 4-foot x 4-foot pan. A 3-inch piece was clipped off the 6-inch tab sticking out of each corner; the remaining 3-inch tab was bent around the corner up against the adjoining side. Two small holes were drilled through the overlapping tab and adjoining side. Small copper harness rivets, the ones with washers on them, were put in those holes and hammered down. When the Boss soldered each corner, top to bottom, he had finished his boiling-down pan.

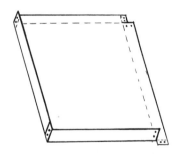

Figure 5-5   Evaporating pan, two sides folded up.

The stone fireplace was another thing all built when I went to work for the Boss. His skill as a stone-mason was obvious: even the 10-foot-high chimney had been constructed without any mortar. The fireplace had two long, low stone walls on which we set the two boiling-down pans. These walls were about 2 feet wide at the top and about 2 feet high. They were 8 feet long and about 3½ feet apart. A 10-foot-tall chimney was at one end with the other end open. There was also an opening at the back side of the chimney so the 4-foot-long cordwood could be fed from this end too. Heavy iron sheets at both ends were used to control the draft. If we wanted more heat we lifted the iron sheet up off the ground, propping it with a flat stone. To cool the fire, we closed it, all much the same as dampers on a woodstove. A sketch of this fireplace is shown in figure 5-6.

* * *

The sugar bush on the Boss's farm grew on a south slope, the best location for sugaring in New England. The maple grove of some 20 to 30 trees, most of them 2 or 3 feet in diameter, was cleared of other types of trees.

Figure 5-6.   A masonry fireplace that can be used to boil down sap.

33

The Boss tended the maples like a garden, spreading the cold ashes from the boiling-down fireplace around them for fertilizer. He said the use of ashes increased both the sap and the sugar in the sap. Might be a good idea if brush being cleared from a sugar bush were put through a hogger and strewn around in the bush as a mulch. Perhaps a little nitrogen (urea, ammonium nitrate or dried blood) should be added to this to help the decay along and get a good humus.

It's not good to pasture animals in a sugar bush. Most of them enjoy eating the sugar maple saplings, making it difficult for you to grow new trees to replace the old or damaged ones.

All the trees in a sugar bush should be tapped once and tested for amounts of sap and sugar. The low-sugar trees could be thinned out to make room for the others. Thinning could also be based on the amount of foliage the tree has. Those with a larger crown tend to yield more sap.

The sap runs best on clear, warm days with sharp, cold frosts at night. This weather pattern usually begins sometime between the last of February and the middle of March, about four weeks, until the buds begin to swell on the tapped trees.

In the old days, the Boss said, people used to get sap from maple trees by using a saw or axe to cut partway into the trunks of the trees, or by breaking off several branches and collecting the sap that dripped from them. Killed plenty of trees, too, he noted.

It was a warm sunny day, that spring forty years ago when the Boss walked down to check the test holes he'd drilled a few days earlier. When the sap was running good, it was time to tap. We fetched a couple of ½-inch augers from the workshop and filled a sap bucket with the sumac spouts. We took these and as many sap buckets as we could carry and headed for the sugar bush.

Using one of the hand augers, the Boss showed me how to drill a tap hole. He turned the auger till it was in to a mark he'd painted 3 inches from the end of the bit, slanting the hole slightly upward. A bit and brace, or a breast drill, would have been easier to use, but the Boss wasn't ready for that much progress, figuring a little hard work never hurt anyone. We knew where to tap each tree, because we had set out flat stones the previous fall to put the sap buckets on. All we had to do was drill a hole about 2 feet above these stones.

The Boss planned to put one tap in a 12-inch tree (about 3 feet around) and he added one more tap for each additional 6 inches in diameter. The tap holes were usually about a foot apart, carefully planned so not to take too much sap from any one tree, which can kill or weaken the tree. The Boss also always tried to tap on the side where the branches were the thickest, usually the side that gets the most sun. He claimed more sap ran on that side.

We had an easy way to put spouts in the tap holes, using a 5-inch piece of ¾-inch pipe with a cap screwed on a threaded end. The Boss put the tapered end of the spout in the tap hole, then put this pipe over the other end. Two or three good licks with a wooden mallet and the spout was in. Sap was dripping out the end shortly after.

The next step was to clear away any snow from the top of the flat stone and set a sap bucket on it with a cover. Aside from keeping out the weather, the cover kept the sap cooler, helping it to stay sweet.

We set out about 20 buckets that spring and averaged about 200 gallons of sap a day for a little over a month. From this, we made 40 gallons of syrup and 250 pounds of sugar — enough for all year for our place with some to spare for the relatives who lived in town.

We carried the buckets of sap from the trees to the fireplace by hand, and the foot or so of snow we had that year didn't make it very easy. Perhaps using a small shoulder yoke to help carry the buckets and snowshoes might have made the task a little easier.

All the boiling down was done over the wood fire. We used hardwood — hickory, maple, or oak — cut into four-foot lengths. We always cut the wood for sugaring the year before, storing it under a canvas tarp near the fireplace, so it would be dry when we needed it. We burned about two cords a week, so we generally kept about ten cords on hand.

Each morning, as soon as the sap was steaming, making that peculiar rolling boil, I kept it clean with the skimmer, a long-handled round wooden paddle about a foot in diameter used for removing the gray scum that formed on top.

Every day at noon, the evaporated sap was poured into the pan nearest the chimney, the "back pan." After that, we'd fill the front pan with fresh sap, bank the fire and go for lunch. We'd stoke the fire up again when we returned. Somewhere late in the afternoon we quit pouring sap into the front pan, put all the boiled-down sap into the back pan and took the front pan off the fire. When the concentrated sap was the right temperature, between 5° and 6° above the boiling point of water (which I'll explain later), we'd pour this from the back pan into a 20-gallon copper wash boiler. This was carried to the house to boil down more on the wood cookstove.

Nights, after supper, we finished boiling down the concentrated sap, using a very accurate candy thermometer. The Boss used the temperature of boiling water to calculate all his syrup and sugar temperatures. Water usually boiled somewhere between 209° F. and 211° F. at our location, depending on the barometric pressure at the time.

The sap in the copper boiler was boiled until it was 7° above the temperature of boiling water. This gave us maple syrup, about a 65% sugar solution, weighing about 11 pounds per gallon. We'd pour this hot through a pre-soaked, all-wool flannel filter to remove the sugar sand and then put the filtered syrup into sterilized quart canning jars, and seal them hot.

When we finished sapping, it was clean-up time. We threw away the sumac spouts since they'd split if stored away. We drove plugs made from maple branches into the tap holes since the Boss felt this would

35

keep the tap holes from getting diseased.

We took the sap buckets into the milkhouse, where we rinsed them and scalded them. The Boss had me take them over to the woodshed and put them upside down on a high shelf. We'd scrape off the scale that had formed on the boiling-down pans and scald them too. When we put them upside down on top of some logs in the wagon shed, we had finished another sapping season.

* * *

We made several different maple sugars from the filtered syrup and we could make it anytime during the year from our canned syrup. The temperature we used and the way it was handled after we boiled down the syrup made the difference in the kind of sugar that was produced. I remember making six different kinds.

Large Crystal Sugar

We boiled to about 10° above the boiling point of water, then set this on the back of the stove where it cooled very slowly. The Boss got clusters of crystals, a few as big as an inch in diameter, after several weeks.

Maple Chew

We boiled to about 20° above the boiling point of water, then quickly cooled this to room temperature by pouring it onto a marble slab, which yielded a thick, chewy candy. We had to eat it quickly, though, since, after only a few days it crystallized.

Maple Fudge

We boiled to about 22° above the boiling point of water, then cooled this to 160° F. We stirred this with a wooden spoon for about 10 minutes, until it turned to a light tan color, indicating the crystals were forming. We poured this into a fudge pan and let it set up. This stayed soft and sugary for several months if kept in a cool place.

Maple Taffy

We boiled to about 24° above the boiling point of water, to where the thick syrup made a long thin string in the air when dripped off a spoon. Without letting this cool, we'd take it outside and pour it onto some clean snow. Delicious.

Maple Cream

We boiled to about 26° above the boiling point of water, cooled this rapidly to 70°, stirring until some crystals formed. It had to be poured into a fudge pan fast or it'd set up hard in the kettle. Made right, it melted in your mouth. We made this often and the Boss's wife sometimes added cream and butternuts to it during stirring, to make candy for special company.

Maple Sugar

We boiled to about 32° above the boiling point of water, let it cool to about 200°, then stirred it a little. The more stirring we were able to do, the finer-grained the sugar, but we were careful not to stir it till it became solid. We'd pour it into pans fast and it would set up before you knew it. We made this more often because it kept well.

— *S. Carl Traegde*

Homemade Ice Cream

"**H**YGIENICALLY SPEAKING, THEY cannot be recommended for the final course of a dinner, as cold mixtures reduce the temperature of the stomach, thus retarding digestion until the normal temperature is again reached. But how cooling, refreshing, and nourishing, when properly taken, and of what inestimable value in the sick room!" This lilting advice came from Fannie Merritt Farmer in her 1896 edition of *The Boston Cooking School Cook Book* in reference to ices, ice creams, and other frozen desserts. Just exactly what she meant by "properly taken" is not imparted to the reader, but any ice cream addict might want to take a wild guess, ignoring those sickroom values — unless by way of justification.

Advice like that must have been about as welcome as hearing about the virtues of cod-liver oil. When the family gathered and the temperature soared on summer Sunday afternoons, reservations about the perils of reducing the temperature of the stomach were undoubtedly swept aside as the ice cream freezer, sack of rock salt, fresh ingredients and tub of ice were brought out for the weekly ice cream ritual, one almost as unbreakable as the visit to church earlier in the day.

Even though this summer Sunday ritual has faded along with taking a turn on the crank of the ice cream freezer, ice cream is still marked by the strong loyalty of its devotees, a loyalty with a surprisingly long historical reach.

If it's true that I scream, you scream, we all scream for ice cream, these cries have come up through the ages from some impressive vocal cords. Marco Polo (a devoted sherbet fan), Catherine de Medici, Richard the Lion-hearted, and our own George Washington, who was rumored to have run up some rather astounding ice cream bills during the hot summer months.

In the early days of the colonies and on into the nineteenth century, ice cream was made by agitating a container of sweetened cream in a tub of salt and ice. The ice cream freezer that is still with us today was invented in 1846 by Nancy Johnson, an otherwise obscure figure on the culinary scene. In fact, it may have been her invention that brought ice cream down from its regal, aristocratic pedestal and onto the tongues of the middle class. After the turn of the century, street vendors known as hokey-pokey men peddled their confections to eager young customers, the ice cream cone was invented and shortly thereafter the indelible names of Good Humor, Eskimo Pie, and Howard Johnson crested the horizon of frozen desserts.

It was sometime after this that ice cream took its turn for the worse. Our fellow ice cream addicts, Catherine de Medici and George Washington would scarcely recognize their beloved confection in today's stabilized, emulsified form. The convenient freezer that most of us enjoy today rules out many of the pleasantries of this ancient delicacy. What we call ice cream, those solid bricks in our fiercely efficient freezing compartment, can hardly compete with its nineteenth century counterpart in flavor or refreshment. Temperature, for one thing, is one of the most important parts of ice cream flavor and the common temperature of most freezers, zero, is just too far this side of cold to allow all the subtle flavors to emerge. But aside from temperature, the quality and freshness of the ingredients is the critical factor in the difference between then and now. You can still enjoy "old-time" ice cream, though, by making it yourself in a hand-crank freezer. Freezers much like Nancy Johnson's original are still produced and are widely available today.

PROCEDURE

The cream now sold in supermarkets is not what it once was. Not only that, it is terrifically expensive. The separated cream from raw milk makes incomparable ice cream and is much more economical than its supermarket counterpart. Raw milk is much richer than store-bought and has a much higher vitamin content. Because of all this, recipes that call for "cream" now yield an ice cream more like ice milk, if store-bought cream is used. So, if it's available to you (most dairy farmers will sell it to you during milking hours if you bring your own container), use raw milk and its cream for these recipes. For a medium cream, let the milk stand undisturbed for 24 hours and for a heavy cream, let it stand for 48 hours.

By the same token, all the ingredients you use should be as fresh as you can possibly manage it, especially fruits. A lot of flavor can be lost, for instance, to limp strawberries, bruised and bleeding on the produce shelf or worse still, strawberries from the frozen food department.

Before you start, scald the can and the dasher and if the staves on the tub are dry and loose, soak the tub in the stream or in a larger bucket filled with water. The rock salt — available from most hardware stores — should be at the ready and the ice well cracked (old-timers used a burlap sack and a wooden mallet for this but an old pillow case and the flat side of a hammer do just as well). Make sure the can as well as the ice cream mixture are well chilled before you start churning and be sure you never fill the can more than three-quarters full — somewhere between two-thirds and three-quarters is best — because, like any cream product, there is expansion as you crank. If the can gets overcrowded, the ice cream's creaminess will be forfeited to a granular consistency.

There are three important things to watch when preparing ice cream: the proportion of ice to salt, careful measuring of ingredients, and the rate at which the crank is turned.

- The proportion of ice and salt is generally three to one. I have seen recipes that recommend a proportion as high as eight parts ice to one part salt but I get best results with the three-to-one formula. Amazingly enough, it's not the ice that makes the ice cream freeze, but the salt *water*, the salt effectively "heating" the ice to a colder temperature. As the ice melts away as you crank, you may feel tempted to add more ice but resist the urge. And no water should be drawn off until the ice cream is ready. By the same token, no more salt should be added. The

Figure 6-1.   Although ice cream doesn't have the same strong appeal during the winter months, making it is simpler when you can use snow instead of cracked ice.

salt speeds up the freezing process but it adversely affects the consistency of the ice cream, making it lose its creamy smoothness in favor of a more coarse, granular consistency. For sherbets or ices, however, you do want to use a larger proportion of salt to ice — equal parts is ideal — since a coarser consistency is desired.

- All the ingredients should be carefully measured, especially the sugar: too sweet a mixture delays the freezing process. That goes for the fruit too, because of its high sugar content.

- The crank should be turned slowly and easily for the first five minutes, then picking up speed for the next 10 minutes or so, until the ice cream is solid.

With everything ready to go, pour the prepared ice cream mixture into the can and nestle the can into the tub fitting. Gradually layer the ice and the salt around the can in proper proportions, turning the crank slowly to let it settle. Let the mixture sit in the iced tub for about 5 minutes, to chill, then begin churning so the can is turning clockwise, making sure everyone takes a turn on the crank. After about 10-15 minutes, the handle will become more and more difficult to turn until suddenly it will resist, which might make you think you've jammed the gears or that the freezer's broken, but that tug means "it's ice cream."

Making sure the top of the can is wiped clean of ice and salt water, check to make sure it's ready. At this point, you may not be able to resist digging in to this silky goodness right away. But if you can restrain yourself, let the ice cream "ripen": remove the dasher and pack the ice cream down into the can with the back of a long-handled spoon. Put the cover back on tightly, putting a cork in the hole where the dasher fit through. Replace the can in the tub, pack it in four parts ice and one part salt, then protect the tub with a thick covering — an old piece of carpet is traditional or a blanket folded several times — leaving it in a cool, shady spot for at least two hours.

Figure 6-2.

This ice cream is nearly ready. As it gets harder to turn the crank, you should check the ice cream's consistency, making sure to clean the salt off the can before opening it up.

If there is ice cream left over after the feast (doubtful), pack it in plastic freezer containers and store in your freezer — though don't expect the flavor to be the same the second time around. Dispose of the salt water very carefully: this highly concentrated saline solution is deadly to plant life and corrosive to your plumbing. Wash the tub and can thoroughly and store upside down in a clean, dry place.

THE RECIPES

Because of the importance of using fresh, in-season ingredients, the recipes included here are selected on a seasonal basis.

Spring — Strawberry Ice Cream

2 cups fresh, ripe strawberries
juice of ½ a lemon
½ cup sugar
¾ cup fresh-squeezed orange
  juice
1 pt. heavy cream, half-beaten
¼ tsp. salt

Pick over the strawberries, using only the best and chop coarsely. Add to this the fruit juices, sugar and salt. Set in a warm place for ½ an hour, stirring once in a while. Add the cream and chill. Freeze in hand freezer, then pack and let ripen for 2 hours.

Summer — Peach Ice Cream

¾ cup fresh peaches, mashed
½ cup honey
juice of ½ a lemon
1½ cups heavy cream, half-beaten

Mix together honey, lemon juice and peaches. Add to beaten cream and put in hand freezer and freeze. Pack and let ripen for 2 hours.

Fall — Cider Sherbet

4 cups fresh-pressed cider
½ cup sugar
1 cup fresh-squeezed orange juice
juice of 2 lemons

Simmer the cider and sugar together for 5 minutes. Cool. Add juices and freeze in hand freezer, using equal parts salt and ice. Eat right away. Makes ½ gallon.

Winter — Chestnut Ice Cream

3 cups milk, scalded
1½ cups sugar
½ tsp. salt
5 egg yolks
¼ cup maple syrup
1 pint cream, half-beaten
1½ cups boiled chestnuts, pureed

Scald the milk and pour over a mixture of the sugar and egg yolks very slowly, stirring all the while. Cook in the top of the double boiler till thickened. Strain through a piece of cheesecloth and cool. Add cream, maple syrup and cooled chestnuts and put into hand freezer. If there's snow on the ground, use that in the freezing process rather than crushing ice. Pack and let ripen for 3 hours. Makes ½ gallon.

Early Spring — Maple Nut Ice Cream

1 cup B-grade maple syrup
1½ quarts cream, half-beaten
1 tbsp. vanilla
¼ tsp. salt

Grade A and Fancy maple syrup are highly overrated. They are more expensive and the flavor is not as rich, so use B-grade for this recipe if you can. Heat the syrup to boiling and boil for 5 minutes. Take off the heat and cool. Mix together the cream, vanilla and salt and add the cooled syrup. Make sure the mixture is cool before pouring into the freezer. At this point, add a cup of chopped nuts — walnuts, almonds, or butternuts. Freeze, then pack and let ripen for 3 hours. Makes 1 gallon.

Any Season — Banana-Nut Ice Cream

2 cups raw milk or light cream, scalded
4 egg yolks
1/3 cup honey
pinch of salt
1 cup heavy cream
4 very ripe bananas
1 cup peanuts, coarsely chopped
Mix the honey and the salt and add egg yolks. Beat till well blended. Pour scalded milk over this mixture very slowly, beating constantly with a wire whisk. Put this into the top of a double boiler and cook, stirring until the mixture coats the spoon. Chill. Add cream, bananas and nuts and freeze in the hand freezer. Pack and let ripen for 2 hours. Makes 1/2 gallon.

Peppermint-Chocolate Ice Cream

1½ cups raw milk or light cream, scalded
1½ cups finely crushed peppermint candies
½ tsp. peppermint extract
2 tbsp. flour
pinch of salt
2 eggs, room temperature
1½ cups heavy cream, half-beaten
12 oz. chocolate bits
Mix milk, extract and 1 cup of the candies (reserving ½ cup) together, and scald in a double boiler. Mix flour and salt and stir in enough milk to make a smooth paste. Stir into rest of milk in double boiler and continue cooking, stirring until thickened. Then cook, covered, for 10 minutes. Beat the eggs and stir into the milk mixture. Return to the double boiler and cook for 1 minute. Cool. Add cream and freeze in the hand freezer. After 5 minutes of churning, add the ½ cup of peppermints and the chocolate bits, then complete the freezing process. Pack and let ripen for 2 hours. Makes ½ gallon.

— *Edie Clark*

Reclaiming Your Garden Soil

OUR FIRST LAND HAD BEEN PO-tatoed to death; our present farm had been neglected into oblivion. Neither condition — lack of use or overuse — is fatal, and the results of restoring spent or unproductive land are among the greatest of the rewards of farming.

Our land wasn't unusual. Most New England land that was once farm is now fallow, overgrown, or completely returned to the forest whence it came. Land can almost always be reclaimed, but the sooner in the cycle it's caught, the easier it is to restore to tillage.

Most small unused farms follow the pattern of our own. There were still fields that were once tilled, and we could see the rut of the dead furrow that ran the breadth of the cleared land. But the edges were fringed with birch, poplar and brush, and a few small wild cherry trees had sprung up here and there.

There were enough rocks barely showing to make one wonder how the fields were ever plowed.

Clumps of grass, goldenrod and milkweed impeded our way as we walked the land. Not that we needed it as proof, but an old harrow stood rusting in the middle of our back field. The woods were full of brush and fallen trees, and blackberry thickets and sweet fern grew where nothing else would.

Details will vary. There may be plowed but worn-out fields, over-grown orchards, washed-out sand banks, or other remnants of the farm that was.

The first job is to cut brush and small trees back to the fence lines. Even if you can't do anything else right away, do this before these trees get the soil acclimated for the pine cycle that will follow. Each en-croaching bush and tree is a part of the cycle and prepares the soil for the next stage. Catching it before the soil has changed significantly is half the battle.

Using a well-made, heavy-duty pair of lopping shears, cut small growth straight across and as close to the ground as possible. A sharp-cut sapling stub will go through a tractor tire. Larger sapling and tree stumps will have to be pulled out.

Then walk the field and note the location of any rocks. You may wonder how they ever plowed the field with so many of them. For one thing, all the rocks weren't there; they heave up with freezing and thawing each year. The larger rocks were probably just plowed around. You may opt for the same solution, but it's best to remove as many of them as you can.

Having cleared the obstacles you are now ready to turn over the land and put in rye or another green manure crop which you will plow back in. At risk of sounding pedantic, let me mention the definition of manure, which has been obscured by incorrect usage. Manure is any organic material which is added to the land to improve the soil. Enriching crops, compost and animal refuse are all manure.

Rye is the best-known green manure crop. Others that enrich the soil include cowpeas, mustard, oats, alfalfa, clover, winter beans and timothy.

Winter rye is good to plant in the fall and plow in two to three weeks before spring planting. For this reason, it is the best choice for the area where your vegetable garden will go. It won't do miracles, but it beats just letting the soil lie there.

The legumes return nitrogen to the soil along with organic material, and are a good choice for long-term soil improvement. White clover is also good for bees if you let it blossom before plowing under. Alfalfa is expensive to plant, but its deep roots will do wonders for soil. Trefoil is a good choice for wet areas.

Cowpeas, mung beans and mustard are good for spring planting. They germinate in cold soil and are planted just as soon as the ground thaws. In four to six weeks they will be about six inches high and can be plowed under. These are good for preparing your vegetable garden if you didn't get to your land in time for fall planting.

Allow two or three weeks between plowing under and planting. If you can, disc harrow the plants twice. A rear-tined roto-tiller will also chop up the vegetation well as it incorporates it into the soil.

The principle of a green manure crop is that as it decays after being plowed under, it returns to the soil all the nutrients it used while growing, plus carbon (and in the case of legumes, nitrogen). Alfalfa is so deep-rooted that it brings nutrients from deep in the soil that would not be available to more shallow-rooted vegetable crops.

In addition to nutrients, the decaying vegetation adds vital organic matter. All types of soil, sand to clay, respond to this treatment and are improved by the organic additions. Water retention and soil quality are both improved.

The return of organic material to the earth isn't a one-time project. It must be continuous in the form of planting or fertilizing with compost, leaves or animal manure, if the decay process is to continue. It is the process itself, as much as the new material added to the soil, which counts, for it is during the decaying process that nutrients are released.

Once you have fertilized your fields with plantings, animal wastes, and other organic matter (each year's mulch plowed under helps, as do shredded leaves), you can further improve it by rotation planting.

This means dividing your land into several areas and planting different things, changing them each year. Alfalfa, corn and wheat are good choices to rotate. Even if you don't use all the crops for food or forage, your soil will be improving instead of deteriorating.

It is usually of first importance to improve the soil where your vegetable garden will be. If you are starting with unbroken ground, the hardest part of this may be finding someone who will plow it, since no one likes to break ground.

Lime the area well and have it harrowed after plowing. Plant winter rye if it's fall or a spring crop if it's spring. Plow these crops in with well-rotted animal manure or good quality compost, or both, and wait three weeks before planting.

That's about all you can do the first year. Repeated each year, with special nutrients added as soil testing indicates they are needed, this process will turn even solid clay or sand into a fine garden in about five or six years.

If that seems like forever, don't despair. That doesn't mean you have to wait that long to harvest your vegetables. Most gardens are grown under less than optimum conditions, and the results are still satisfying. Your garden will be easier to care for and more productive each year.

If you have plenty of space and plan to keep chickens or other fowl, you can fence off twice as much as you need for the garden and alternate its use between garden and chicken yard — one end for each. This has one serious drawback, however, since most people (including me) want their vegetable garden close to the house and their chickens as far away from it as possible!

Whether your land improvement program is a long-range one or a crash program to improve your tomato crop, you won't do anything on your farm that will bring more lasting satisfaction than leaving your land better than you found it.

— *Barbara Radcliffe Rogers*

45

Building Bridges

A HUNDRED YEARS AGO BRIDGE builders like John Roebling and his son Washington were regarded with as much esteem as we now regard aerospace engineers — and, viewed in the perspective of its time, the functions were similar: bridges make areas of land available that are otherwise inaccessible. When the Brooklyn Bridge was opened in 1883, people were actually trampled to death in the fervor of celebration. Brooklyn is now a section of New York City so well integrated into the mesh of the metropolis that it hardly seems possible that Brooklyn could once have been considered remote or outlying. But, for want of a bridge, it was.

In much the same way, building a bridge can be one of the most practical homesteading necessities. It can open up parts of your land that might otherwise lie useless or it can make a piece of land you might be

considering buying much more attractive by virtue of the potential of its increased accessibility.

In high-level application, acute engineering skills are necessary to successfully span a given space. But building a bridge is neither as complicated nor as simple as one might imagine. Two types of bridges are outlined here: a footbridge that can weather spring floods and be sturdy enough for you and your wheelbarrow and a beam bridge, suitable for you and your garden tractor, but not sturdy enough for cars and trucks.

THE FOOTBRIDGE

The footbridge will do nicely if you are simply interested in getting across a stream or run-off area to a prospective garden patch or small pasture. This is a clever old design, well worth reviving, for crossing streams that frequently swell over their banks in the spring. It is an unattached bridge, designed harmoniously with these natural spring occurrences: the bridge moves along with the rising floodwater, but is held from floating downstream by four posts, one in each corner of the bridge (see figure 8-1).

Because of its free form, this bridge can be built in your garage or basement. You'll need two hemlock, spruce, or fir (rough-cut, if possible) 4x4s for the stringers, the supporting members that reach from one side of the stream bank to the other. Thoroughly soak the timbers in creosote with special attention to the endgrain, the part most susceptible to rot. Pressure-treated wood is best but it's really too expensive for this design. Logs could also be used instead of the 4x4s, but be sure to remove the bark and soak them in creosote. Also, split off the top quarter to give a flat nailing surface. The

length of the stringers will be the width of your span plus a foot extra on either side. The planks for the deck should be well creosoted 2x4s. It's unlikely that you'll want the bridge to be wider than 3 or 4 feet, so the planks should be that width. Position the planks so 4 or 5 inches hang over each stringer and fasten them, after drilling holes for the 20d galvanized nails. Leave a ½- or ¾-inch space between each plank to ensure air circulation and prevent moisture from being trapped between the planks (see figure 8-2).

When you are securing the planks to the stringers, skip the next to the last plank on either end, leaving space for the four posts. Finish off the deck with one last plank, flush with the ends of the stringers. The posts should be higher than the highwater mark — between six inches and a foot should be more than adequate — and should be driven into the ground at all four corners of the bridge.

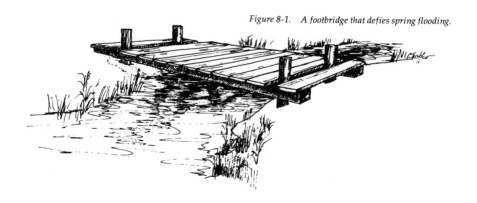

Figure 8-1. A footbridge that defies spring flooding.

To ensure the longest life for your footbridge, put flat rocks or pieces of slate under each of the stringers resting on the banks. Also, take the bridge up every few years and give it a fresh soaking of creosote, top and bottom.

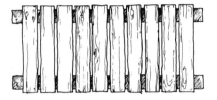

Figure 8-2. Space the planks on the deck ½ or ¾ of an inch apart to discourage rot.

THE BEAM BRIDGE

The beam bridge is more complicated and takes more time but produces a much sturdier, more serviceable bridge. With proper weight load calculations, you could even make a bridge of this sort bear loads like cars and trucks — but you're on your own for that.

The stringers for this bridge will be hemlock, spruce, or fir 6x10s and the deck members will be 2x6s, which will be quite adequate for you and your tractor, plus a trailer loaded with topsoil or cordwood.

Like all building foundations, the most important part of the structure of the beam bridge are the abutments, the structural supports on either side of the stream's banks. These masonry abutments, being strong enough to withstand the flow of the stream, will already be stronger than necessary to hold up the stringers, the deck and the load crossing the bridge. Ideally, this bridge should not have piers, or midway supports, so the span should not be more than 16 feet. Stone or concrete are the best materials to use for the abutments and stone will probably be free for the taking somewhere in or near your stream.

These abutments must be arced, arching end to end toward the stream banks, convex toward the stream. This prevents the stream from channeling around behind the abutments, thus eroding the stream banks and weakening this strongest part of your bridge. Also, the arc cooperates with the action of the stream, rather than impeding it, as would be the case with a square abutment, at right angles to the flow of the stream (see figure 8-3).

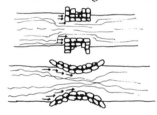

Figure 8-3. Abutments should be made in an arc. If they are at right angles to the stream's flow, they are weakened by the constant push of the water and are more likely to be washed away by floodwaters.

The best time to begin construction of the bridge is in the dry season, September or October, when the level of your stream is at its lowest. Even if you get a late start on it and can't get it finished until spring, you're way ahead with the abutments completed.

However, even during the dry season, you still have to divert the flow of water away from your project. Build a cofferdam, a temporary enclosure that will keep the water out of your way. Make it of wood, metal or plastic, favoring seamless, nonporous materials like heavy plastic sheeting or used aluminum offset printing plates (in some areas these are available from the local newspaper for a nominal fee), and burrow it into the stream bed. Technically a cofferdam is watertight but chances are that no matter how good an enclosure you rig up, it's still not going to be perfectly watertight and some water will be leaching in around your work area.

As you would for any good foundation or stone structure, you have to have a footing. Dig down below the stream bed a good foot. This is a little like shoveling against the tide, so work quickly and have a good supply of the biggest rocks you can handle on hand. When you get the trench dug, nestle these large stones tightly and snugly until they rise up above the level of the stream bed.

Then make up a *dry* Portland concrete mix, calculating the area you need to fill as if you were making a wet batch of concrete. Pour this dry mix in around the rocks. The very moist conditions you're working under along with the leaching stream water will adequately mix with the dry mortar to form a good strong bond.

From there, build your stonework up, laying the mortar and stones in as true an arc as you can.

When the abutments are nearly to the level of the banks, leave a notch in either end (see figure 8-4) large

Figure 8-4.

enough to accommodate the stringers. The notches ought to be deep enough so that the top part of the stringer will rise slightly above the top of the abutments, so that your deck will be level.

All the lumber used for this bridge must be thoroughly creosoted. Rot is a bridge's worst enemy and there's nothing quite so disheartening as going to all the work of building a structure such as this and having it rot out a few years later. Pressure-treated wood is best and if you can afford it, it's worth it for this bridge. In its stead, give the stringers and the deck members several soakings of creosote, with special attention to the endgrain, the most susceptible part. Also, if it's available in your area, rough-cut lumber bought right at the sawmill can save you money over that that's been planed.

So, all things being level and timbers armed against the elements, lay the timbers deadweight in the notches. Make sure they don't hang beyond the notches since this will put them in contact with the soil and encourage rot — in spite of the creosote. Using 20d galvanized nails, fasten the 2x6 deck members to the stringers, leaving about a ½-

49

Figure 8-6.   The completed beam bridge is as aesthetically appealing as it is functional.

or ¾-inch space between each plank so air can circulate and keep the planks dry against rain and flooding.

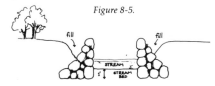

Figure 8-5.

You will most likely need fill to fill in the space created between the abutments and the stream banks (see figure 8-5). This fill should be graded gradually toward the bridge to make a gentle approach, making it not only easy on your tractor but also less threatening for horses or mules, whose aversion to crossing bodies of water is infamous.

You may want to add "bumpers" on either side of the bridge, which can be done simply by nailing down 2x4s the length of your span, flush with the outside of the bridge. A railing can also be added but keep in mind that a railing will limit the width of what can be brought over the bridge. Without a railing, you can haul broadside loads.

If properly constructed and maintained, you can count on either or both of these bridges to last for many years to come — a small price to pay for making otherwise inaccessible land useful.

— *Edie Clark*

Flax to Linen

LINEN. TODAY THAT WORD MEANS mostly sheets, towels and table-cloths. But not so very long ago linen meant simply any woven material made from flax. Around the time of the founding of this country almost every New England household had a new supply of the fabric made from home-grown, spun and woven flax. The quality table and bed linen, and the linen clothing, were the treasures of the household and were passed lovingly from generation to generation.

But then in the early 1800s when many settlements were just getting a good start, the industrial revolution made cheap cotton material readily available. Why go to all the trouble of growing and processing flax if a good substitute could be had for pennies? The tradition of growing and processing flax for linen faded from most memories.

Mary Chase of South Brooksville,

Maine, is trying, almost single-handedly, to revive that memory. She has grown small patches of flax at her home for several years now, and recently started to process it by hand into spun flax for weaving. She has imported seed from Holland and Sweden and sold some to about 25 people in Maine, and even some to the historic community of Cooperstown, New York.

Her overwhelming enthusiasm for the fabric is contagious as she displays some of the lovely and intricately woven pieces she has collected. The hand-woven material has a natural luster that makes it look like polished metal, and ranges in color from off-white through silver gray to a light brown.

Mrs. Chase learned the techniques of handling flax from the only source she could find that offered extensive training in the subject — a group of handcraft people in

51

Sweden. A weaver by profession, Mrs. Chase took a six-month course in weaving in Sweden in 1972, including some flax work, and later returned for a one-week course in flax culture and preparation conducted by Henryk Zinkievicz, an agronomist with the Swedish Seed Growing Association.

The flax plant needs about a three-month growing season, and generally can be raised anywhere grain is grown. It requires some moisture throughout the growing season and does best on a medium loam, but one that is not too rich. Flax grown on soils high in nitrogen often grows too fast, is coarse and has a low fiber content.

When sown thickly in mid- or late May on a smooth firm seed bed, the plants will be ready to harvest in late August or early September. A small patch of around 20 feet by 30 feet sown with two pounds of flax-seed should provide enough fiber to weave a small tablecloth or a towel.

The seeds are planted close together to encourage the plants to grow tall, and not branch until the very top. When harvested the plants should be about three feet high, and ideally no more than one or two millimeters in diameter. Mrs. Chase cultivates a special variety of flax (*Linum usitatissimum* which means "most useful") grown for its fiber, and not the type grown commercially in this country for its seed from which linseed oil is pressed.

Harvesting flax and preparing it for the spinning wheel is a long and laborious process, which is probably one of the reasons the art was so readily given up when a substitute for linen became available. "It's hard and dirty work," says Mrs. Chase.

The plants are ready to harvest about 90 days after sowing, when the lower third of the stalks have yellowed and withered, but before the seeds have fully ripened. Most of the plant is still green at this point.

The three-foot plants are pulled up by their short shallow roots, never cut. Then they are dried by tying them in bundles and hanging them over a line or on a rack. From the harvesting to the spinning, it is important to keep the flax in neat bundles with the root ends butted together. "Otherwise you have a mess," says Mrs. Chase.

When the plants are dry, the seeds are removed with the aid of a ripple which looks like an oversized comb. The iron or wooden comb is stationary and the bundles of flax are drawn through it a handful at a time (see figure 10-1). The round seed pods which look like small dried berries pop off and should be collected for the next planting, or for use as animal feed or pressed for the oil. The ripple Mrs. Chase uses was made by a local woodworker. She said she has used a fine-toothed garden rake as a ripple with some success.

The de-seeded plants are then carefully laid out on the ground or submerged in a pond to "ret." This is a decomposition process that dissolves the gums that bind the flax fibers to the woody portion of the stem and to each other.

In dew retting, the flax plants are laid out parallel on the grassy ground, one layer deep. Halfway

Figure 10-1. Rippling flax. In the window in the background, processed flax in various stages — the large, cone-shaped bundle on the left is flax arranged on a distaff and tied with a ribbon, ready to be spun into thread.

through the process the plants are carefully flipped over end to end with a broom or rake handle so that both sides of the plants have contact with the ground. In pond retting, the bundles of flax are tied to small rafts and weighted with stones to keep them submerged. Mrs. Chase remembers reading that cattle were not allowed to drink from ponds where flax was retting. "I suppose that was because it is a rotting process, and ideally the ponds are almost stagnant," neither exactly healthy conditions for drinking water.

Flax can be retted in a deep vat indoors, where a small amount of water is drawn off and replaced with fresh water every day. The time required for the retting process depends on the water temperature.

Dew-retted flax is attacked by a fungus responsible for the decomposition, while bacteria are the agents in the pond retting. Water-retted flax is lighter in color, but dew-retted flax is easier to "scutch," which is the process of removing the decomposed woody fibers.

Mrs. Chase is dew-retting her flax, a process that must be watched closely for two or three weeks and tested periodically to make sure the outer covering has decomposed, yet not rotted the prized linen fibers.

To test the retting process a sample flax plant is brought indoors to dry, and then bent and broken along its length to see if the outer cuticle and woody pith will break and separate from the lustrous linen fibers. When they separate readily, they're ready for the next step.

After the retting, the stalks are again dried, and then a handful at a time are broken with a "brake" which looks like an overgrown school paper cutter with several dull wooden blades (see figure 10-2). The process doesn't harm the strong and flexible linen fibers of the plant, but breaks the retted woody portion of the stalk into pieces that can be scraped, or scutched, off. Men on the olden farms often carried out the breaking, the hardest part of the process. If a brake wasn't available, farmers would beat the daylights out of the stalks with a wooden mallet to achieve the same results.

Figure 10-3. Scutching broken flax against a scutching block with a carved wooden scutching knife.

Figure 10-2. Mrs. Mary Chase breaking retted flax with a brake.

Scutching, or scraping off the woody pieces of stalk, is done against a scutching block with a scutching knife (see figure 10-3). The block is a vertical board over which a handful of the broken flax is draped. Then a large wooden knife is worked down the flax on the block scraping off the broken pieces of woody tissue. One of the knives Mrs. Chase uses was made by her son Andy.

After scutching, the bundle of flax fiber is drawn through a hackle or hetchel, to straighten the long fibers out and to comb out the short fibers known as "tow" (see figure 10-4). The hackle is a block set with spikes that looks like a small section of an Indian nailbed, and brings to mind the old saying "don't get your hackles up." The word "towhead" is derived from the short coarse tow fibers combed out by the hackle. Ideally the flax is drawn through a series of hackles with progressively finer teeth.

Tow, being strong but short, coarse and rough, was spun and used for making candlewicks, twine, cord or rope, or woven for heavy sacking or work shirts. The finer, three-foot-long flax fibers were spun and woven into the treasured linen.

After hackling, the bundles of flax are carefully arranged around a special long spire for easier spinning into thread. These spires are called "distaffs" and may be simple boards or very ornate miniature towers like the pair of two-foot-high hand-carved ones Mrs. Chase got in

Figure 10-4. Hackling.

Sweden. When the flax is on the distaff, it looks like a big wad of cotton candy. A ribbon is wrapped around it to keep it in place while the spinner draws individual fibers from the bundle with moistened fingers and spins them into linen thread on a spinning wheel.

It has been discovered that saliva is better than water to moisten the fibers while they are being spun. Apparently saliva makes the natural gums in the flax stick to each other resulting in a better thread. Flax has the potential for turning into a very fine thread, depending on the skill of the spinner. And the finer the thread, the larger the material which may be woven from a given quantity of flax.

Although flax doesn't have a natural twist to the fiber the way cotton does, which would make it easier to spin, it does have the advantage of being one of man's longest natural fibers with a staple of up to 90 cm or about ten times the staple length of cotton. That length is the main reason linen is practically lint free, making a piece of it ideal for washing windows or mirrors. Its fibers have been demonstrated to be as strong as steel of the same thickness, and linen is even stronger when wet than dry. It's very absorbent, which makes clothing or sheets made of linen always seem so cool even in the hottest weather.

But the natural luster is the quality which makes linen attractive to most people. A table set with linen cloth and napkins fairly gleams. Old linen shirts take on the sheen of a polished suit of armor. This quality seems to be enhanced as the material is used, washed, bleached by the sun and pressed with a cool iron.

Flax fibers have a 5,000-year history, having been found in Swiss lake dwelling of that long ago. But Mrs. Chase notes that "there isn't a strong tradition of linen in this country. It was more a utilitarian skill" and lasted only as long as the need for the skill existed.

Mrs. Chase has a copy of a letter written in 1876 by Mrs. Jonah Holt, an ancestor of Roland Howard of nearby Blue Hill, Maine. The letter, written at the time of this country's centennial celebration, describes the flax process as Mrs. Holt remembered her grandparents doing it. It's almost identical to the process Mrs. Chase is following. But such finds are rare. "I would be enormously grateful for any information from anyone on flax grown and cultured in New England," Mrs. Chase said. She would also love to see any handwoven linen anyone may have as heirlooms.

With the durability of the fabric what it is, there's a good chance some of Mrs. Chase's seven children will have family heirlooms of their own to pass on to their grandchildren.

— *Jean Heavrin*

Herbal Medicines:
The Garden Apothecary

OLD HERBALS ABOUND IN "SIMPLES" and "specifics" for every manner of disorder from broken bones to itchy skin. But it was the common discomfort — colds, coughs, scratches, sore muscles, bruises, insect bites, and general aches and pains — that every housewife knew how to relieve with plants she grew in her dooryard and dried from the rafters.

In a day when there wasn't a drugstore on every corner, a medicine shelf in every grocery store, and a clinic only a telephone call away, it was very important to know how to relieve those afflictions not serious enough to require riding out to summon a doctor.

Those who scoff at home remedies today in the age of miracle drugs, and call herbal medicines "witch's brews," forget that the herbs our grandmothers administered are the basis for many of our current remedies. Vitamin C tablets for colds, wild cherry and horehound coughdrops, and peppermint-flavored antacids are nothing more than old herbal medicines gone commercial.

The housewife of old didn't know why they worked, but modern medicine has made some fascinating discoveries which uphold many of the remedies she knew. Spring greens, considered a much-needed tonic after a winter without fresh vegetables, were an important source of vitamins A and C.

Jewelweed, long known by rural dwellers as an antidote to poison ivy, has been found by University of Vermont researchers to contain anti-fungal agents. Several commercial preparations for treating ivy poisoning contain extracts of this plant.

Antiseptic, astringent, diuretic, stimulant, sedative — all these medicinal qualities belong to one or several of the common herbs. They are used externally and internally, and are prepared in teas, decoctions, lotions, compresses, cough syrups and baths.

Herbal teas, or tisanes, are the most common. They are prepared much as China teas are. Dried or fresh herbs are steeped in boiling water, usually in a warmed china teapot or cup (never in metal) so that the heat is retained while the tea infuses. The aroma and flavor of the herbs are released in about ten minutes of steeping.

The amounts suggested vary from one teaspoon to one tablespoon for a cup of water. It is better to use more of the herb than to try steeping it longer, since the herbs often develop a bitter flavor when steeped too long.

Jewelweed

A decoction is a medicine made by boiling the herb (often seeds, roots and barks are treated in this way) in water, usually for about five minutes. The flavor is stronger and often bitter. Certain herbs, such as lemon balm, require decocting even for a tea in order to release their essential oils. Some herb teas are gently

stimulating, as is China tea, but unlike it, contain no caffeine.

A number of herb teas or tisanes are drunk for their calming and relaxing effect. Chamomile, fennel, dill, catnip and lemon verbena are the most commonly named.

Peter Rabbit's mother gave him chamomile tea after a rather trying day, and although literature doesn't report it, Mrs. MacGregor probably gave her husband some to soothe his nerves over the loss of his radish crop. Those with pollen allergies, however, should not drink it.

Hops, a vine which is frequently found around old farmhouses, has a well documented sleep-producing ingredient, lupulin, and is often made into hops pillows.

Anise seed steeped in milk is supposed to be a sleep-producing drink, but it is also quite likely that the warm milk alone would accomplish the purpose.

For those who can't drink coffee for a morning lift, there are herbal stimulants. Hyssop, peppermint, spearmint and savory are all good pick-me-up teas.

Lemonbalm

Hops

Peppermint is known for its ability to ease a queasy stomach, and for good reason. It is a digestant and is used as such in a number of patent preparations. Sage and anise teas are also frequently mentioned as aids to digestion.

Peppermint is often taken boiled in milk. The Shakers used feverfew steeped in hot water to relieve stomachache, and in fancier circles, candied caraway seeds were passed after dinner as comfits to aid digestion.

Chamomile

Decoctions of anise seed, fennel, coriander and caraway and teas of catnip and sweet marjoram were recommended for colic. The aroma of anise is very strong in a number of modern prescriptions for colic. All of these seeds, like the modern preparations, leave a warm sensation in the stomach, which is comforting to a colicky child.

The smell of lavender is said to cure a headache. Rosemary, which is also mentioned as a sedative, is frequently suggested as a cure for headache. Some suggest it as a tea, others using it as a compress.

There is frequent disagreement in old herbals not only as to what the herb remedied, but how to use it. Sweet fern is usually agreed upon as an antidote for poison ivy, but

Peppermint

sources vary on whether to drink it or bathe in it. So it is with mints, chamomile, and rosemary; so if you don't like them as a tea, you can save them for your bathwater.

Almost any herb can be added to the bath for fragrance, but a number have qualities which make them medicinal as well. Calendula (pot marigold) and comfrey are both

Rosemary

soothing to the skin, and pennyroyal is good for itching. Calendula juice was used during World War I for its wound-healing properties and comfrey contains allantoin, which acoounts for its soothing effect.

Pennyroyal is an insect repellent which keeps away mosquitoes and black flies if rubbed on exposed skin. It will also relieve the itch afterwards if you forget to use it first. It should never be taken internally.

Since the common cold has plagued man for centuries, as it will probably continue to do for centuries yet to come, it is not surprising that every early cookbook contains receipts for cures or at least relief from its symptoms.

Horehound

One of the most frequent is to keep the patient warm and induce perspiration. A number of herbs will do this when drunk as very hot teas. Hyssop, thyme, lemon balm and chamomile are mentioned. If you are short of herbs, hot lemonade is a good substitute. Modern medicine would indicate that lemonade, rose hip tea or a tea made

of wild strawberry leaves might be most effective because of their high vitamin C content.

For the sore throat and cough that come with the cold, wild cherry and horehound is made with a cup of dried leaves and blossoms boiled in two cups of water for ten minutes.

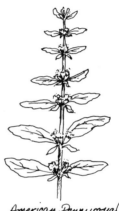

American Pennyroyal

Let it steep an additional ten minutes and strain. If it is too strong, add boiling water. As a tea, it will ease a tickling throat or a cough, but is even more effective when one part of the infusion is mixed with two parts honey.

To make your own cough drops, add two cups of sugar to one of infusion and two pinches of cream of tartar. Stir over low heat until the sugar has dissolved and cook until it reaches the hard ball stage (290°) on a candy thermometer. Pour on a buttered platter or marble slab and mark off into small squares. When it cools and hardens, crack it along these lines to make cough drops.

Sage tea mixed with honey will also aid a sore throat, as will thyme tea. (Oil of thyme is the base of at least one patent cough medicine.) Both elderberry and black currant jelly are often suggested as soothing for a cough or tickle in the throat, but any jelly or seedless jam will have the same effect, as does plain corn syrup or honey.

Compresses and poultices, along with mustard plasters, haven't been out of style quite so long as herb teas, and will probably take longer to make a comeback. There is good reason for this. Few who endured a mustard plaster as a child will want to repeat the experience.

But herbal remedies include a number of these and they are neither as messy nor as complicated as they sound. Compresses are easily made by steeping the dried herb in very little boiling water, straining the mixture through a cloth, which is then folded or tied to contain the wet leaves.

Chamomile, a versatile medicinal herb, is suggested as a compress for tired eyes (and the resulting tea can be used as a soothing eyewash). Comfrey is recommended as a compress to reduce swelling and soothe skin irritations.

Caraway seeds, rosemary and costmary are all suggested to relieve pain from bruises, and a compress of hyssop leaves will prevent black and blue marks if applied immediately to potential bruises or shiners. Oregano, which has a slight numbing quality, makes a soothing poultice for aching muscles.

Insect bites and bee stings have long plagued those who live in the country, and in addition to pennyroyal, there are a number of herbs that are reported to relieve the effects of these. For bee sting, try rubbing with leaves of fresh hyssop, savory or lemon balm.

A compress of hot parsley leaves is said to relieve itching of insect bites, but a piece of cotton dipped in *very* hot water and held on the bite until the sting subsides (a few seconds) will dispel the same itch for hours and you can save the parsley for your salad, where it will fortify you with vitamins A and C and iron.

There are many others. Interesting sources of additional home

Winter Savory

remedies, herbal and others, are old cookbooks. This kitchen medicine was just as important to the young housewife as knowing how to prepare and preserve food.

The only problem with these old remedies is that they often relieve conditions we no longer have — or at least not by the same names. Cures abound for tetters, chilblains, scurvy, ague, erysipelas, and summer diseases. Enough maladies remain, however, that everyone should be able to come up with a cure — or two or three — for his favorite symptom.

These homey remedies may or may not relieve your distress. They are not a substitute for medical care for those who are ill, but they may save some discomfort in the same way that you would seek help from common patent medicines. In any case, they won't hurt anyone, they might help, and they will certainly prove an interesting diversion to help take your mind off your misery.

— *Barbara Radcliffe Rogers*